IS-700.a - National Incident Management System (NIMS), An Introduction

By FEMA

Based on Public Domain Text

IS-700.A - National Incident Management System (NIMS),

An Introduction

Course Overview

This course provides an introduction to the National Incident Management System (NIMS).

Objectives: At the end of this course, you will be able to:

- Describe the intent of NIMS.

- Describe the key concepts and principles underlying NIMS.

- Describe the purpose of the NIMS Components including: Preparedness, Communications and Information Management, Resource Management, and Command and Management.

- Describe the purpose of the National Integration Center.

This course provides a basic introduction to NIMS. It is not designed to replace Incident Command System and position-specific training.

Lesson 1: Understanding NIMS

What Is NIMS?

Each day communities respond to numerous emergencies. Most often, these incidents are managed effectively at the local level.

However, there are some incidents that may require a collaborative approach that includes personnel from:

- Multiple jurisdictions,
- A combination of specialties or disciplines,
- Several levels of government,
- Nongovernmental organizations, and
- The private sector.

The National Incident Management System, or NIMS, provides the foundation needed to ensure that we can work together when our communities and the Nation need us the most.

NIMS integrates best practices into a comprehensive, standardized framework that is flexible enough to be applicable across the full spectrum of potential incidents, regardless of cause, size, location, or complexity.

Using NIMS allows us to work together to prepare for, prevent, respond to, recover from, and mitigate the effects of incidents.

This course introduces you to the NIMS concepts, principles, and components.

National Incident Management System (NIMS) Overview

NIMS provides a consistent nationwide template to enable Federal, State, tribal, and local governments, nongovernmental organizations, and the private sector to work together to prevent, protect against, respond to, recover from, and mitigate the effects of incidents regardless of cause, size, location, or complexity in order to reduce the loss of life and property and harm to the environment.

NIMS Document: A Collaborative Partnership

The NIMS document was developed through a collaborative intergovernmental partnership with significant input from the incident management functional disciplines, NGOs, and the private sector.

Originally published on March 1, 2004, the NIMS document was revised in 2008 to reflect contributions from stakeholders and lessons learned during recent incidents.

Related NIMS Document Section

This lesson summarizes the information presented in the Introduction and Overview, including:

- Introduction

- Concepts and Principles

    o Flexibility

    o Standardization

- Overview of NIMS Components

HSPD-5, Management of Domestic Incidents

Homeland Security Presidential Directive (HSPD) 5, "Management of Domestic Incidents," directed the Secretary of Homeland Security to:

- Develop and administer a National Incident Management System (NIMS).

- Develop the National Response Framework (NRF).

The NIMS Mandate

HSPD-5 requires all Federal departments and agencies to:

- Adopt NIMS and use it in their individual incident management programs and activities.

- Make adoption of NIMS by State, tribal, and local organizations a condition for Federal preparedness assistance (through grants, contracts, and other activities).

Collaborative Incident Management

NIMS is **not** an operational incident management or resource allocation plan.

NIMS represents a core set of doctrines, concepts, principles, terminology, and organizational processes that enables effective, efficient, and collaborative incident management.

NIMS Builds on Best Practices

Building on the foundation provided by existing emergency management and incident response systems used by jurisdictions, organizations, and functional disciplines at all levels, NIMS integrates best practices into a comprehensive framework.

These best practices lay the groundwork for the components of NIMS and provide the mechanisms for the further development and refinement of supporting national standards, guidelines, protocols, systems, and technologies.

NIMS Is Dynamic

NIMS is **not** a static system.

NIMS fosters the development of specialized technologies that facilitate emergency management and incident response activities and allows for the adoption of new approaches that will enable continuous refinement of the system over time.

NIMS Components

NIMS is much more than just using the Incident Command System or an organization chart.

NIMS is a consistent, nationwide, systematic approach that includes the following components:

- Preparedness

- Communications and Information Management

- Resource Management

- Command and Management

- Ongoing Management and Maintenance

The components of NIMS were not designed to stand alone, but to work together.

Preparedness

Effective emergency management and incident response activities begin with a host of preparedness activities conducted on an ongoing basis, in advance of any potential incident. Preparedness involves an integrated combination of assessment; planning; procedures and protocols; training and exercises; personnel qualifications, licensure, and certification; equipment certification; and evaluation and revision.

Communications and Information Management

Emergency management and incident response activities rely on communications and information systems that provide a common operating picture to all command and coordination sites. NIMS describes the requirements necessary for a standardized framework for communications and emphasizes the need for a common operating picture. This component is based on the concepts of interoperability, reliability, scalability, and portability, as well as the resiliency and redundancy of communications and information systems.

Resource Management

Resources (such as personnel, equipment, or supplies) are needed to support critical incident objectives. The flow of resources must be fluid and adaptable to the requirements of the incident. NIMS defines standardized mechanisms and establishes the resource management process to identify requirements, order and acquire, mobilize, track and report, recover and demobilize, reimburse, and inventory resources.

Command and Management

The Command and Management component of NIMS is designed to enable effective and efficient incident management and coordination by providing a flexible, standardized incident management structure. The structure is based on three key organizational constructs: the Incident Command System, Multiagency Coordination Systems, and Public Information.

Ongoing Management and Maintenance

Within the auspices of Ongoing Management and Maintenance, there are two components: the National Integration Center (NIC) and Supporting Technologies.

Flexibility

The components of NIMS are adaptable and scalable to any situation, from routine, local incidents, to incidents requiring the activation of interstate mutual aid, to those requiring a coordinated Federal response. NIMS applies to all types of incidents.

Standardization

NIMS provides a set of **standardized** organizational structures that improve integration and connectivity among jurisdictions and disciplines, starting with a common foundation of preparedness and planning.

Personnel and organizations that have adopted the common NIMS framework are able to work together, thereby fostering cohesion among the various organizations involved in all aspects of an incident.

What Is NIMS?

What NIMS is:

- A comprehensive, nationwide, systematic approach to incident management, including the Incident Command System, Multiagency Coordination Systems, and Public Information

- A set of preparedness concepts and principles for all hazards

- Essential principles for a common operating picture and interoperability of communications and information management

- Standardized resource management procedures that enable coordination among different jurisdictions or organizations

- Scalable so it may be used for all incidents (from day-to-day to large-scale)

- A dynamic system that promotes ongoing management and maintenance

What NIMS is NOT:

- A response plan

- Only used during large-scale incidents

- A communications plan

- Only applicable to certain emergency management/incident response personnel

- Only the Incident Command System or an organization chart

- A static system

Lesson 2: NIMS Preparedness

Related NIMS Document Section

This lesson summarizes the information presented in Component I: Preparedness, including:

- Concepts and Principles

- Achieving Preparedness

What Is NIMS Preparedness?

Given the threats we face, a lack of preparedness could have catastrophic consequences. Effective and coordinated emergency management and incident response require that we create a culture of preparedness.

National preparedness can only succeed through coordination at all levels of government and by forming strong partnerships with the private sector and nongovernmental organizations.

Preparation is a continuous cycle of planning, organizing, training, equipping, exercising, evaluating, and taking corrective action.

NIMS provides the mechanisms and tools to help enhance preparedness. Within NIMS, preparedness focuses on:

- Planning,

- Procedures and protocols,

- Training and exercises,

- Personnel qualification and certification, and

- Equipment certification.

The concepts and principles that form the basis for preparedness are the integration of the concepts and principles of all the components of NIMS.

This lesson introduces you to the NIMS Preparedness component.

NIMS and Other Preparedness Efforts

Homeland Security Presidential Directive (HSPD) 5 established a single, comprehensive approach to incident management. The following additional Homeland Security Presidential Directives are linked to national preparedness:

- **HSPD-7: Critical Infrastructure Identification, Prioritization, and Protection** established the U.S. policy for "enhancing protection of the Nation's critical infrastructure and key resources" and mandates a national plan to implement that policy in partnership with Federal departments and agencies; State, tribal, and local governments; nongovernmental organizations; and the private sector.

- **HSPD-8: National Preparedness** directed DHS to lead a national initiative to develop a National Preparedness System—a common, unified approach to "strengthen the preparedness of the United States to prevent and respond to threatened or actual domestic terrorist attacks, major disasters, and other emergencies."

NIMS and the National Response Framework

The National Response Framework (NRF):

- Is a guide to how the Nation conducts all-hazards response.

- Builds upon the NIMS coordinating structures to align key roles and responsibilities across the Nation, linking all levels of government, nongovernmental organizations, and the private sector.

A basic premise of both NIMS and the NRF is that incidents typically are managed at the local level first. Following NIMS doctrine, the NRF is designed to ensure that local jurisdictions retain command, control, and authority over response activities for their jurisdictional areas.

Elected and Appointed Officials

To better serve their constituents, elected and appointed officials must understand and commit to NIMS.

NIMS provides elected and appointed officials with a framework to help:

- Ensure agency/jurisdiction policies for emergency management and incident response are clearly stated.

- Evaluate effectiveness and correct any deficiencies.

- Support a coordinated, multiagency approach.

Although elected and appointed officials may not be at the scene of the incident, they should have the ability to communicate and support the on-scene command.

Elected and appointed officials should have a clear understanding of their roles and responsibilities for successful emergency management and incident response. These officials include administrative and political personnel as well as department/agency administrators who have leadership roles in a jurisdiction, including legislators and chief executives, whether elected (e.g., Governors, mayors, sheriffs, tribal leaders, and county executives) or appointed (e.g., county administrators and city managers). Although their roles may require providing direction and guidance to constituents during an incident, their day-to-day activities do not necessarily focus on emergency management and incident response.

To better serve their constituents, elected and appointed officials should do the following:

- Understand, commit to, and receive training on NIMS and participate in exercises.

- Maintain an understanding of basic emergency management, continuity of operations/continuity of government plans, jurisdictional response capabilities, and initiation of disaster declarations.

- Lead and encourage preparedness efforts within the community, agencies of the jurisdiction, nongovernmental organizations (NGOs), and the private sector, as appropriate.

- Help to establish relationships (including mutual aid agreements and assistance agreements) with other jurisdictions and, as appropriate, with NGOs and the private sector.

- Support and encourage participation in mitigation efforts within the jurisdiction and, as appropriate, with NGOs and the private sector.

- Provide guidance to their jurisdictions, departments, and/or agencies, with clearly stated policies for NIMS implementation.

- Understand laws and regulations in their jurisdictions that pertain to emergency management and incident response.

- Maintain awareness of critical infrastructure and key resources within their jurisdictions, potential incident impacts, and restoration priorities.

Elected and appointed officials may also be called upon to help shape and revise laws, policies, and budgets to aid in preparedness efforts and to improve emergency management and incident response activities.

An incident may have a mix of political, economic, social, environmental, public safety, public health, and financial implications with potentially serious long-term effects. Frequently, incidents require a coordinated response (across agencies, jurisdictions, and/or including NGOs and the private sector), during which elected and appointed officials must make difficult decisions under crisis conditions. Elected and appointed officials should be aware of how NIMS can work to ensure cooperative response efforts, thereby minimizing the potential implications of an incident.

Preparedness: Continuous Cycle

Ongoing preparedness helps us to:

- Coordinate during times of crisis.

- Execute efficient and effective emergency management and incident response activities.

Preparedness is achieved and maintained through a continuous cycle of planning, organizing, training, equipping, exercising, evaluating, and taking corrective action.

Preparedness: A Unified Approach

Preparedness requires a unified approach to emergency management and incident response activities. To achieve a unified approach, components of NIMS should be integrated within the emergency management and incident response structure.

Preparedness should be integrated into resource management, command and management, and communications and information management to form an effective system.

Levels of Capability

For NIMS to function effectively, jurisdictions and organizations should set expectations about the capabilities and resources that will be provided before, during, and after an incident.

Inventorying and categorizing of resources is a critical element of preparedness because it:

- Establishes and verifies the levels of capability needed based on risk and hazard assessments prior to an incident.

- Identifies and verifies that emergency response resources possess the needed qualifications during an incident.

Coordination of Preparedness Activities

Preparedness activities should be coordinated among all appropriate agencies and organizations within the jurisdiction, as well as across jurisdictions. Preparedness activities may involve the following groups:

- Individuals

- Preparedness Organizations

- Nongovernmental Organizations

- Private Sector

Individuals

Individuals should participate in their community's outreach programs that promote and support individual and community preparedness (e.g., public education, training sessions, demonstrations). These programs should include preparedness of those with special needs.

Preparedness Organizations

Preparedness organizations provide coordination for emergency management and incident response activities before an incident or scheduled event.

These organizations range from groups of individuals to small committees to large standing organizations that represent a wide variety of committees, planning groups, and other organizations (e.g., Citizen Corps, Local Emergency Planning Committees, Critical Infrastructure Sector Coordinating Councils).

Nongovernmental Organizations

Nongovernmental organizations (NGOs), such as community-based, faith-based, or national organizations (e.g., the Salvation Army, National Voluntary Organizations Active in Disaster, and the American Red Cross), play vital support roles in emergency management and incident response activities.

Compliance with NIMS is not mandated for nongovernmental organizations. However, adherence to NIMS can help these organizations integrate into a jurisdiction's preparedness efforts.

To ensure integration, capable and interested nongovernmental organizations should be included in ongoing preparedness efforts, especially in planning, training, and exercises.

Private Sector

The private sector plays a vital support role in emergency management and incident response and should be incorporated into all aspects of NIMS. Utilities, industries, corporations, businesses, and professional and trade associations typically are involved in critical aspects of emergency management and incident response.

These organizations should prepare for all-hazards incidents that may affect their ability to deliver goods and services. It is essential that private-sector organizations that are directly involved in emergency management and incident response (e.g., hospitals, utilities, and critical infrastructure owners and operators) be included in a jurisdiction's preparedness efforts, as appropriate.

Governments at all levels should work with the private sector to establish a common set of expectations consistent with Federal, State, tribal, and local roles, responsibilities, and methods of operations. These expectations should be widely disseminated and the necessary training and practical exercises conducted so that they are thoroughly understood in advance of an actual incident.

NIMS Preparedness Efforts

Preparedness efforts should validate and maintain plans, policies, and procedures, describing how they will prioritize, coordinate, manage, and support information and resources. This section of the lesson describes the following preparedness efforts:

- Planning

- Procedures & Protocols

- Training & Exercises

- Personnel Qualifications & Certification

- Equipment Certification

Continuity Capability

Recent natural and manmade disasters have demonstrated the need for building continuity capability as part of preparedness efforts. Continuity planning should be instituted within all organizations (including all levels of government and the private sector) and address such things as:

- Essential functions.

- Orders of succession.

- Delegations of authority.

- Continuity facilities.

- Continuity communications.

- Vital records management.

- Human capital.

NSPD-51/HSPD-20 and Federal Continuity Directive 1, dated February 4, 2007, outline the continuity requirements for all Federal departments and agencies (with guidance for non-Federal organizations).

Mutual Aid Agreements and Assistance Agreements

Mutual aid agreements and assistance agreements provide a mechanism to quickly obtain emergency assistance in the form of personnel, equipment, materials, and other associated services.

NIMS encourages:

- Jurisdictions to enter into mutual aid and assistance agreements with other jurisdictions and/or organizations from which they expect to receive, or to which they expect to provide, assistance.

- States to participate in interstate compacts and to consider establishing intrastate agreements that encompass all local jurisdictions.

Types of Mutual Aid Agreements and Assistance Agreements

There are several types of these kinds of agreements, including but not limited to the following:

- Automatic Mutual Aid

- Local Mutual Aid

- Regional Mutual Aid

- Statewide/Intrastate Mutual Aid

- Interstate Agreements

- International Agreements

- Other Agreements

Automatic Mutual Aid
Agreements that permit the automatic dispatch and response of requested

resources without incident-specific approvals. These agreements are usually basic contracts; some may be informal accords.

Local Mutual Aid
Agreements between neighboring jurisdictions or organizations that involve a formal request for assistance and generally cover a larger geographic area than automatic mutual aid.

Regional Mutual Aid
Substate regional mutual aid agreements between multiple jurisdictions that are often sponsored by a council of governments or a similar regional body.

Statewide/Intrastate Mutual Aid
Agreements, often coordinated through the State, that incorporate both State and local governmental and nongovernmental assets in an attempt to increase preparedness statewide.

Interstate Agreements
Out-of-State assistance through formal State-to-State agreements such as the Emergency Management Assistance Compact, or other formal State-to-State agreements that support the response effort.

International Agreements
Agreements between the United States and other nations for the the exchange of Federal assets in an emergency.

Other Agreements
Any agreement, whether formal or informal, used to request or provide assistance and/or resources among jurisdictions at any level of government (including foreign), NGOs, or the private sector.

Procedural Documents

Effective preparedness involves documenting specific procedures to follow before, during, and after an incident.

Procedural documents should detail the specific actions to implement a plan or system. There are four standard levels of procedural documents:

- Standard Operating Procedure or Operations Manual

- Field Operations Guide or Incident Management Handbook

- Mobilization Guide

- Job Aid

Standard Operating Procedure or Operations Manual
Complete reference document that provides the purpose, authorities, duration,

and details for the preferred method of performing a single function or a number of interrelated functions in a uniform manner.

Field Operations Guide or Incident Management Handbook
Durable pocket or desk guide that contains essential information required to perform specific assignments or functions.

Mobilization Guide
Reference document used by agencies/organizations outlining agreements, processes, and procedures used by all participating organizations for activating, assembling, and transporting resources.

Job Aid
Checklist or other visual aid intended to ensure that specific steps for completing a task or assignment are accomplished. Job aids serve as training aids to teach individuals how to complete specific job tasks.

Protocols

Protocols are sets of established guidelines for actions (which may be designated by individuals, teams, functions, or capabilities) under various specified conditions.

Establishing protocols provides for the standing orders, authorizations, and delegations necessary to permit the rapid execution of a task, function, or a number of interrelated functions without having to seek permission.

Protocols permit specific personnel—based on training and delegation of authority—to assess a situation, take immediate steps to intervene, and escalate their efforts to a specific level before further guidance or authorizations are required.

Training

Personnel with roles in emergency management and incident response should be appropriately trained to improve all-hazards capabilities nationwide. Training should allow practitioners to:

- Use the concepts and principles of NIMS in exercises, planned events, and actual incidents.

- Become more comfortable using NIMS, including the Incident Command System.

Training and exercises should be specifically tailored to the responsibilities of the personnel involved in incident management. The National Integration Center (NIC) has developed requirements and guidance for NIMS training materials.

Exercises

To improve NIMS performance, emergency management/response personnel need to participate in realistic exercises. Exercises should:

- Include multidisciplinary, multijurisdictional incidents.

- Require interactions with the private sector and nongovernmental organizations.

- Cover all aspects of preparedness plans, particularly the processes and procedures for activating local, intrastate, and/or interstate mutual aid agreements and assistance agreements.

- Contain a mechanism for incorporating corrective actions and lessons learned from incidents into the planning process.

Personnel Qualifications and Certification

A critical element of NIMS preparedness is the use of national standards that allow for common or compatible structures for the qualification, licensure, and certification of emergency management/response personnel. Standards:

- Help ensure that personnel possess the minimum knowledge, skills, and experience necessary to execute incident management and emergency response activities safely and effectively.

- Typically include training, experience, credentialing, validation, and physical and medical fitness.

The baseline criteria for voluntary credentialing will be established by the National Integration Center.

Equipment Certification

We all count on having the right tools to do the job. Being able to certify equipment is a critical component of preparedness. Equipment certification:

- Helps ensure that the equipment acquired will perform to certain standards (as designated by organizations such as the National Fire Protection Association or National Institute of Standards and Technology).

- Supports planning and rapid fulfillment of needs based on a common understanding of the abilities of distinct types of equipment.

Mitigation and Preparedness

Mitigation is an important element of emergency management and incident response. Mitigation:

- Provides a critical foundation in the effort to reduce the loss of life and property and to minimize damage to the environment from natural or manmade disasters by avoiding or lessening the impact of a disaster.

- Provides value to the public by creating safer communities and impeding the cycle of disaster damage, reconstruction, and repeated damage. These activities or actions, in most cases, will have a long-term sustained effect.

Preparedness planning and mitigation planning are complementary processes that should support one another.

Mitigation Activities

Risk management—the process for measuring or assessing risk and developing strategies to manage it—is an essential aspect of mitigation. Risk management strategies may include avoiding the risk (e.g., removing structures in floodplains), reducing the negative effect of the risk (e.g., hardening buildings by placing barriers around them), or accepting some or all of the consequences of a particular risk.

Examples of mitigation activities include the following:

- Ongoing public education and outreach activities designed to reduce loss of life and destruction of property.

- Complying with or exceeding floodplain management and land-use regulations.

- Enforcing stringent building codes, seismic design standards, and wind-bracing requirements for new construction, or repairing or retrofitting existing buildings.

- Supporting measures to ensure the protection and resilience of critical infrastructure and key resources designed to ensure business continuity and the economic stability of communities.

- Acquiring damaged homes or businesses in flood-prone areas, relocating the structures, and returning the property to open space, wetlands, or recreational uses.

- Identifying, utilizing, and refurbishing shelters and safe rooms to help protect people in their homes, public buildings, and schools in hurricane- and tornado-prone areas.

- Implementing a vital records program at all levels of government to prevent loss of crucial documents and records.

- Intelligence sharing and linkage leading to other law enforcement activities, such as infiltration of a terrorist cell to prevent an attack.

- Periodic remapping of hazard or potential hazard zones, using geospatial techniques.

- Management of data regarding historical incidents to support strategic planning and analysis.

- Development of hazard-specific evacuation routes.

Lesson 3: NIMS Communications and Information Management

Related NIMS Document Section

This lesson summarizes the information presented in Component II: Communications and Information Management, including:

- Concepts and Principles

- Management Characteristics

- Organization and Operations

What Is NIMS Communications and Information Management?

Effective emergency response depends on communication—the ability to maintain a common operating picture through the constant flow of information.

During and after Hurricane Katrina, communications systems failed, severely hampering information flow and response operations. In New Orleans, most of the city was flooded. The combined effects of wind, rain, storm surge, breached levees, and flooding knocked out virtually the entire infrastructure—electrical power, roads, water supply and sewage, and communications systems.

Thomas Stone, Fire Chief, St. Bernard Parish: "We lost our communications system, and when you are not able to communicate, you can't coordinate your response. You never think that you will lose your entire infrastructure."

Communications problems are not limited to systems being destroyed or not functioning. Similar problems arise when agencies cannot exchange needed information because of incompatible systems. NIMS identifies several important features of public safety communications and information systems.

Communications systems need to be . . .

- **Interoperable**—able to communicate within and across agencies and jurisdictions.

- **Reliable**—able to function in the context of any kind of emergency.

- **Portable**—built on standardized radio technologies, protocols, and frequencies.

- **Scalable**—suitable for use on a small or large scale as the needs of the incident dictate.

- **Resilient**—able to perform despite damaged or lost infrastructure.

- **Redundant**—able to use alternate communications methods when primary systems go out.

Regardless of the communications hardware being used, standardized procedures, protocols, and formats are necessary to gather, collate, synthesize, and disseminate incident information. And in a crisis, life-and-death decisions depend on the information we receive.

This lesson introduces you to the NIMS Communications and Information Management component.

Flexible Communications and Information Systems

All too often, after-action reports cite communications failures as an impediment to effective incident management.

Communications breakdowns are not limited to equipment and systems-related failures. The use of different protocols, codes instead of plain language, and nonstandardized reporting formats hampers our ability to share critical information and make effective decisions.

To overcome these past problems, the NIMS Communications and Information Management component promotes the use of flexible communications and information systems.

Common Operating Picture

A common operating picture is established and maintained by gathering, collating, synthesizing, and disseminating incident information to all appropriate parties.

Achieving a common operating picture allows on-scene and off-scene personnel—such as those at the Incident Command Post, Emergency Operations Center, or within a Multiagency Coordination Group—to have the same information about the incident, including the availability and location of resources and the status of assistance requests.

Interoperability

First and foremost, **interoperability is the ability of emergency management/response personnel to interact and work well together.**

Interoperability also means that technical emergency communications systems should:

- Be the same or linked to the same system that the jurisdiction uses for nonemergency procedures.

- Effectively interface with national standards, as they are developed.

- Allow the sharing of data throughout the incident management process and among all key players.

Interoperability Saves Lives!

Jan. 13, 1982: Air Florida Flight 90 crashed into the 14th St. Bridge in Washington, DC, during a snowstorm. More than 70 people lost their lives. Police, fire, and EMS crews responded quickly to the scene but experienced coordination problems because they could not communicate with one another.

Sept. 11, 2001: When American Airlines Flight 77 crashed into the Pentagon, 900 responders from 50 different agencies were able to communicate with one another. Response agencies had learned an invaluable lesson from the Air Florida tragedy. Regional coordination within the Washington area led to the adoption of the Incident Command System, establishment of interoperable communications protocols, and execution of mutual aid plans. The next challenge to solve was the lack of direct interoperability with secondary response agencies.

Reliability, Portability, and Scalability

To achieve interoperability, communications and information systems should be designed to be:

- **Reliable**—able to function in any type of incident, regardless of cause, size, location, or complexity.

- **Portable**—built on standardized radio technologies, protocols, and frequencies that allow communications systems to be deployed to different locations and integrated seamlessly with other systems.

- **Scalable**—suitable for use on a small or large scale, allowing for an increasing number of users.

Resiliency and Redundancy

Communications systems ensure that the flow of information will not be interrupted during an incident through:

- **Resiliency**—able to withstand and continue to perform after damage or loss of infrastructure.

- **Redundancy**—providing for either duplication of identical services or the ability to communicate through diverse, alternative methods when standard capabilities suffer damage.

Standardized Communications Types

Successful communications and information management require that emergency management/response personnel and their affiliated organizations use the following types of standardized communications:

- Strategic Communications

- Tactical Communications

- Support Communications

- Public Address Communications

The determination of the individual or agency/organization responsible for these communications is discussed in the NIMS Command and Management lesson.

Strategic Communications: High-level directions, including resource priority decisions, roles and responsibilities determinations, and overall incident response courses of action.

Tactical Communications: Communications between command and support elements and, as appropriate, cooperating agencies and organizations.

Support Communications: Coordination in support of strategic and tactical communications (for example, communications among hospitals concerning resource ordering, dispatching, and tracking from logistics centers; traffic and public works communications).

Public Address Communications: Emergency alerts and warnings, press conferences, etc.

Policy and Planning

Coordinated communications policy and planning provides the basis for effective communications and information management. Based on policies, communications plans should include procedures and protocols that identify:

WHAT	What information is essential. What information can be shared.
WHO	Who needs the information. Who has the information.
HOW	How information will flow among all stakeholders (including the private sector, critical infrastructure owners and operators, and nongovernmental organizations). How information is coordinated for release to the public and media. How communications systems and platforms will be used (including technical parameters of all equipment and systems).

All relevant stakeholders should be involved in planning sessions in order to formulate integrated communications plans and strategies. Technology and equipment standards also should be shared when appropriate, to provide stakeholders with the opportunity to be interoperable and compatible.

Policy and Planning: Guidelines

Sound communications management policies and plans should include information about the following aspects of communications and information management:

- Information needs should be defined by the jurisdiction/organization. These needs are often met at the Federal, State, tribal, and local levels, in concert with NGOs and the private sector, and primarily through preparedness organizations.

- The jurisdiction's or organization's information management system should provide guidance, standards, and tools to enable the integration of information needs into a common operating picture when needed.

- Procedures and protocols for the release of warnings, incident notifications, public communications, and other critical information are disseminated through a defined combination of networks used by the

Emergency Operations Center. Notifications are made to the appropriate jurisdictional levels and to NGOs and the private sector through defined mechanisms specified in emergency operations and incident action plans.

- Agencies at all levels should plan in advance for the effective and efficient use of information management technologies (e.g., computers, networks, and information-sharing mechanisms) to integrate all command, coordination, and support functions involved in incident management and to enable the sharing of critical information and the cataloging of required corrective actions.

Agreements

Agreements should be executed among all stakeholders to ensure that the elements within plans and procedures will be in effect at the time of an incident.

Agreements should specify all of the communications systems and platforms through which the parties agree to use or share information.

Equipment Standards and Training

Standards help ensure a seamless interface between communications systems, especially between the public and private sectors. Standards should address:

- The wide range of conditions under which communications systems must operate.

- The need for maintenance and updating of the systems and equipment.

- The periodic testing of systems.

Periodic training and exercises are essential so that personnel capabilities and limitations of communications plans and systems are addressed before an incident.

Incident Information

Shared information is vital to the Incident Commander, Unified Command, and decisionmakers within supporting agencies and organizations. A single piece of information may provide input for:

- Development of incident objectives and the Incident Action Plan (IAP).

- Identification of safety hazards.

- Determination of resource needs.

- Formulation of public information messages.

- Analysis of incident costs.

Examples of Incident Information

The following are examples of information generated by an incident that can be used for decisionmaking purposes:

- Incident Notification, Situation, and Status Reports

- Analytical Data

- Geospatial Information

Incident Notification, Situation, and Status Reports

Incident reporting and documentation procedures should be standardized to ensure that situational awareness is maintained and that emergency management/response personnel have easy access to critical information. Situation reports offer a snapshot of the past operational period and contain confirmed or verified information regarding the explicit details (who, what, when, where, and how) relating to the incident. Status reports, which may be contained in situation reports, relay information specifically related to the status of resources (e.g., availability or assignment of resources).

The information contained in incident notification, situation, and status reports must be standardized in order to facilitate its processing; however, the standardization must not prevent the collection or dissemination of information unique to a reporting organization. Transmission of data in a common format enables the passing of pertinent information to appropriate jurisdictions and organizations and to a national system that can handle data queries and information/intelligence assessments and analysis.

Analytical Data

Data, such as information on public health and environmental monitoring, should be collected in a manner that observes standard data collection techniques and definitions. The data should then be transmitted using standardized analysis processes. During incidents that require public health and environmental sampling, multiple organizations at different levels of government often collect data, so standardization of data collection and analysis is critical. Additionally, standardization of sampling and data collection enables more reliable analysis and improves the quality of assessments provided to decisionmakers.

Geospatial Information

Geospatial information is defined as information pertaining to the geographic location and characteristics of natural or constructed features and boundaries. It is often used to integrate assessments, situation reports, and incident notification into a common operating picture and as a data fusion and analysis tool to synthesize many kinds and sources of data and imagery. The use of geospatial data (and the recognition of its intelligence capabilities) is increasingly important during incidents. Geospatial information capabilities (such as nationally consistent grid systems or global positioning systems based on lines of longitude and latitude) should be managed through preparedness efforts and integrated within the command, coordination, and support elements of an incident, including resource management and public information.

The use of geospatial data should be tied to consistent standards, as it has the potential to be misinterpreted, transposed incorrectly, or otherwise misapplied, causing inconspicuous yet serious errors. Standards covering geospatial information should also enable systems to be used in remote field locations or devastated areas where telecommunications may not be capable of handling large images or may be limited in terms of computing hardware.

Communications and Data Standards

Communications and data standards are established to allow diverse organizations to work together effectively. Standards may include:

- A standard set of organizational structures and responsibilities.

- Common "typing" of communications resources to reflect specific capabilities.

- Use of agreed-upon communications protocols.

- Common identifier "titles" for personnel, facilities, and operational locations used to support incident operations.

Plain Language and Common Terminology

The **use of plain language** in emergency management and incident response:

- Is a matter of safety.

- Facilitates interoperability across agencies/organizations, jurisdictions, and disciplines.

- Ensures that information dissemination is timely, clear, acknowledged, and understood by all intended recipients.

Codes should not be used, and all communications should be confined to essential messages. The use of acronyms should be avoided during incidents requiring the participation of multiple agencies or organizations.

Encryption or Tactical Language

When necessary, information may need to be encrypted so that security can be maintained.

Although plain language may be appropriate during response to most incidents, tactical language is occasionally warranted due to the nature of the incident (e.g., during an ongoing terrorist event).

The protocols for using specialized encryption and tactical language should be incorporated into the Incident Action Plan or incident management communications plan.

Public Information

Providing effective incident information to the public is an important element of incident management.

- The **Joint Information System (JIS)** integrates incident information and public affairs into a cohesive organization designed to provide consistent, coordinated, accurate, accessible, and timely information.

- The **Joint Information Center (JIC)** provides a structure for developing and delivering incident-related coordinated messages by developing, recommending, and executing public information plans and strategies. The JIC is the central point of contact for all news media at the scene of an incident.

Additional information on these elements is presented in the Command and Management component.

Information Security

Procedures and protocols must be established to ensure information security. Inadequate information security can result in the release of untimely, inappropriate, and piecemeal information that can compound an already complicated situation by:

- Placing responders and community members in danger.

- Increasing the spread of rumors and inaccurate information.

- Disrupting the critical flow of proper information.

- Wasting resources and valuable time correcting the misperceptions.

The release of inappropriate classified or sensitive public health or law enforcement information can jeopardize national security, ongoing investigations, or public health.

Lesson 4: NIMS Resource Management

Related NIMS Document Section

This lesson summarizes the information presented in Component III: Resource Management, including:

- Concepts and Principles

- Managing Resources

What Is NIMS Resource Management?

During an incident, getting the right resources, to the right place, at the right time, can be a matter of life and death.

NIMS establishes a standardized approach for managing resources before, during, and after an incident.

Resources include:

- Personnel,

- Equipment,

- Supplies, and

- Facilities.

Prior to an incident, resources are inventoried and categorized by kind and type, including their size, capacity, capability, skills, and other characteristics.

Mutual aid partners exchange information about resource assets and needs. Resource readiness and credentialing are maintained through periodic training and exercises.

When an incident occurs, standardized procedures are used to:

- Identify resource requirements,

- Order and acquire resources, and

- Mobilize resources.

The purpose of tracking and reporting is accountability. Resource accountability helps ensure responder safety and effective use of incident resources. As incident objectives are reached, resources may no longer be necessary. At this point, the recovery and demobilization process begins.

Recovery may involve the rehabilitation, replenishment, disposal, or retrograding of resources, while demobilization is the orderly, safe, and efficient return of an incident resource to its original location and status. And finally, any agreed-upon reimbursement is made.

When disaster strikes, we must be able to take full advantage of all available and qualified resources. In this lesson you will learn how NIMS provides the mechanisms for ensuring that we can be inclusive and integrate resources from all levels of government, the private sector, and nongovernmental organizations.

Standardized Approach to Resource Management

NIMS establishes a standardized approach for managing resources before, during, and after an incident. This standardized approach is based on the underlying concepts:

- Consistency

- Standardization

- Coordination

- Use

- Information Management

- Credentialing

Consistency: Resource Management provides a **consistent** method for identifying, acquiring, allocating, and tracking resources.

Standardization: Resource Management includes **standardized** systems for classifying resources to improve the effectiveness of mutual aid agreements and assistance agreements.

Coordination: Resource Management includes **coordination** to facilitate the integration of resources for optimal benefit.

Use: Resource Management planning efforts incorporate **use** of all available resources from all levels of government, nongovernmental organizations, and the private sector, where appropriate.

Information Management: Resource Management integrates **communications and information management** elements into its organizations, processes, technologies, and decision support.

Credentialing: Resource Management includes the use of **credentialing** criteria that ensure consistent training, licensure, and certification standards.

Planning

Jurisdictions should work together in advance of an incident to develop plans for identifying, ordering, managing, and employing resources.

The planning process should result in:

- Identification of resource needs based on the threats to and vulnerabilities of the jurisdiction.

- Development of alternative strategies to obtain the needed resources.

- Creation of new policies to encourage positioning of resources.

- Identification of conditions that may trigger a specific action, such as restocking supplies when inventories reach a predetermined minimum.

Use of Agreements

Agreements among all parties providing or requesting resources help to enable effective and efficient resource management during incident operations.

You might want to consider developing and maintaining standing agreements and contracts for services and supplies that may be needed during an incident.

Resource Identification and Ordering

The resource management process uses standardized methods to identify, order, mobilize, and track the resources required to support incident management activities. Identification and ordering of resources are intertwined.

Those with resource management responsibilities perform these tasks either at the request of the Incident Commander or in accordance with planning requirements.

Effective Resource Management: Acquisition Strategies

Effective resource management includes establishing resource acquisition procedures. It is important to consider the tradeoffs (e.g., shelf life, warehousing costs) and determine the optimal acquisition strategies, including:

- Acquiring critical resources in advance and storing them in a warehouse (i.e., "stockpiling").

- Supplying resources "just in time," typically using a preincident contract.

Effective Resource Management: Systems and Protocols

Effective resource management includes:

- **Systems:** Management information systems collect, update, and process resource data and track the status and location of resources.

  It is critical to have redundant information systems or backup systems to manage resources in the event that the primary system is disrupted or unavailable.

- **Protocols:** Preparedness organizations develop standard protocols to request resources, prioritize requests, activate and mobilize resources to incidents, and return resources to normal status.

Managing Resources

The focus of this section of the lesson is on a standardized seven-step cycle for managing resources during an incident.

It is important to remember that preparedness activities must occur on a continual basis to ensure that resources are ready for mobilization.

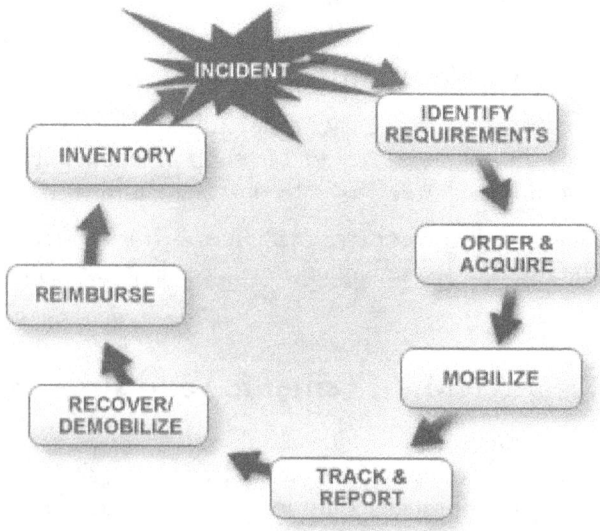

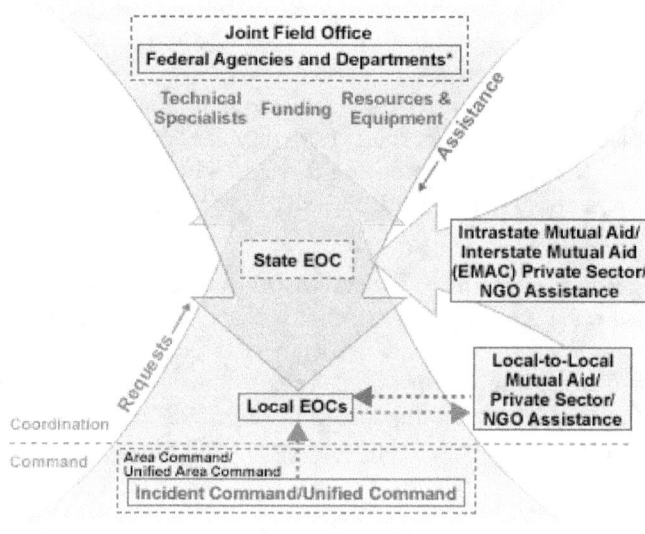

Step 1: Identify Requirements

When an incident occurs, personnel who have resource management responsibilities should continually identify, refine, and validate resource requirements. This process includes identifying:

- What and how much is needed.

- Where and when it is needed.

- Who will be receiving or using it.

Resource availability and requirements constantly change as the incident evolves. Coordination among all response partners should begin as early as possible, preferably prior to incident response activities.

Flow of Requests and Assistance During Large-Scale Incidents

The Incident Command/Unified Command identifies resource requirements and communicates needs through the Area Command (if established) to the local Emergency Operations Center (EOC). The local EOC fulfills the need or requests assistance through mutual aid agreements and assistance agreements with private-sector and nongovernmental organizations.

In most incidents, local resources and local mutual aid and assistance agreements will provide the first line of emergency response and incident management. If the State cannot meet the needs, they may arrange support from another State through an agreement, such as the Emergency Management Assistance Compact (EMAC), or through assistance agreements with nongovernmental organizations.

If additional resources and/or capabilities are required beyond those available through interstate agreements, the Governor may ask the President for Federal assistance.

Federal assistance may be provided under various Federal authorities. If a Governor requests a disaster declaration, the President will consider the entirety of the situation including damage assessments and needs. The President may declare a major disaster (section 401 of the Robert T. Stafford Disaster Relief and Emergency Assistance Act).

The Joint Field Office is used to manage Federal assistance (technical specialists, funding, and resources/equipment) that is made available based on the specifics and magnitude of the incident. In instances when an incident is projected to have catastrophic implications (e.g., a major hurricane or flooding), States and/or the Federal Government may position resources in the anticipated incident area.

In cases where there is time to assess the requirements and plan for a catastrophic incident, the Federal response will be coordinated with State, tribal, and local jurisdictions, and the pre-positioning of Federal assets will be tailored to address the specific situation.

*Note that some Federal agencies (U.S. Coast Guard, Environmental Protection Agency, etc.) have statutory responsibility for response and may coordinate and/or integrate directly with affected jurisdictions.

Step 2: Order & Acquire

Standardized resource-ordering procedures are used when requests for resources cannot be fulfilled locally. Typically, these requests are forwarded first to an adjacent locality or substate region and then to the State.

Decisions about resource allocation are based on organization or agency protocol and possibly the resource demands of other incidents.

Mutual aid and assistance resources will be mobilized only with the consent of the jurisdiction that is being asked to provide the requested resources. Discrepancies between requested resources and those available for delivery must be communicated to the requestor.

Avoid Bypassing Systems

Those responsible for managing resources, including public officials, should recognize that reaching around the official resource coordination process within the Multiagency Coordination System supporting the incident(s) creates serious problems.

Requests from outside the established system can put responders at risk, and at best typically lead to inefficient use and/or lack of accounting of resources.

Step 3: Mobilize

Incident resources mobilize as soon as they are notified through established channels. Mobilization notifications should include:

- The date, time, and place of departure.

- Mode of transportation to the incident.

- Estimated date and time of arrival.

- Reporting location (address, contact name, and phone number).

- Anticipated incident assignment.

- Anticipated duration of deployment.

- Resource order number.

- Incident number.

- Applicable cost and funding codes.

When resources arrive on scene, they must be formally checked in.

Mobilization and Demobilization

Managers should plan and prepare for the demobilization process at the same time that they begin the resource mobilization process.

Early planning for demobilization facilitates accountability and makes the logistical management of resources as efficient as possible—in terms of both costs and time of delivery.

The Demobilization Unit in the Planning Section develops an Incident Demobilization Plan containing specific demobilization instructions.

Step 4: Track & Report

Resource tracking is a standardized, integrated process conducted prior to, during, and after an incident to:

- Provide a clear picture of where resources are located.

- Help staff prepare to receive resources.

- Protect the safety and security of personnel, equipment, and supplies.

- Enable resource coordination and movement.

Resources are tracked using established procedures continuously from mobilization through demobilization.

Step 5: Recover/Demobilize

Recovery involves the final disposition of all resources, including those located at the incident site and at fixed facilities. During this process, resources are rehabilitated, replenished, disposed of, and/or retrograded.

Demobilization is the orderly, safe, and efficient return of an incident resource to its original location and status. As stated earlier, demobilization planning should begin as soon as possible to facilitate accountability of the resources.

During demobilization, the Incident Command and Multiagency Coordination System elements coordinate to prioritize critical resource needs and reassign resources (if necessary).

Nonexpendable Resources

Nonexpendable resources (such as personnel, firetrucks, and durable equipment) are fully accounted for both during the incident and when they are returned to the providing organization. The organization then restores the resources to full functional capability and readies them for the next mobilization. Broken or lost items should be replaced through the appropriate resupply process, by the organization with invoicing responsibility for the incident, or as defined in existing agreements. It is critical that fixed-facility resources also be restored to their full functional capability in order to ensure readiness for the next mobilization. In the case of human resources, such as Incident Management Teams, adequate rest and recuperation time and facilities should be provided. Important occupational health and mental health issues should also be addressed, including monitoring the immediate and long-term effects of the incident (chronic and acute) on emergency management/response personnel.

Expendable Resources

Expendable resources (such as water, food, fuel, and other one-time-use supplies) must be fully accounted for. The incident management organization bears the costs of expendable resources, as authorized in financial agreements executed by preparedness organizations. Restocking occurs at the point from which a resource was issued. Returned resources that are not in restorable condition (whether expendable or nonexpendable) must be declared as excess according to established regulations and policies of the controlling jurisdiction, agency, or organization. Waste management is of special note in the process of recovering resources, as resources that require special handling and disposition (e.g., biological waste and contaminated supplies, debris, and equipment) are handled according to established regulations and policies.

Step 6: Reimburse

Reimbursement provides a mechanism to recoup funds expended for incident-specific activities. Consideration should be given to reimbursement agreements prior to an incident. Processes for reimbursement play an important role in establishing and maintaining the readiness of resources.

Preparedness plans, mutual aid agreements, and assistance agreements should specify reimbursement terms and arrangements for:

- Collecting bills and documentation.

- Validating costs against the scope of the work.

- Ensuring that proper authorities are secured.

- Using proper procedures/forms and accessing any reimbursement software programs.

Step 7: Inventory

Resource management uses various resource inventory systems to assess the availability of assets provided by jurisdictions.

Preparedness organizations should inventory and maintain current data on their available resources. The data are then made available to communications/dispatch centers, Emergency Operations Centers, and other organizations within the Multiagency Coordination System.

Resources identified within an inventory system are not an indication of automatic availability. The jurisdiction and/or owner of the resources has the final determination on availability.

Identifying and Typing Resources

Resource typing is categorizing, by capability, the resources requested, deployed, and used in incidents. The National Integration Center typing protocol provides incident managers the following information:

- **Resource Category:** Identifies the function for which a resource would be most useful.

- **Kind of Resource:** Describes what the resource is (for example: medic, firefighter, Planning Section Chief, helicopter, ambulance, combustible gas indicator, bulldozer).

- **Type of Resource:** Describes the size, capability, and staffing qualifications of a specific kind of resource.

Resource typing must be a continuous process based on measurable standards.

Credentialing

The credentialing process involves an objective evaluation and documentation of an individual's:

- Current certification, license, or degree,

- Training and experience, and

- Competence or proficiency.

Credentialing personnel ensures that they meet nationally accepted standards and are able to perform specific tasks under specific conditions. Credentialing is separate from badging, which takes place at the incident site in order to control access.

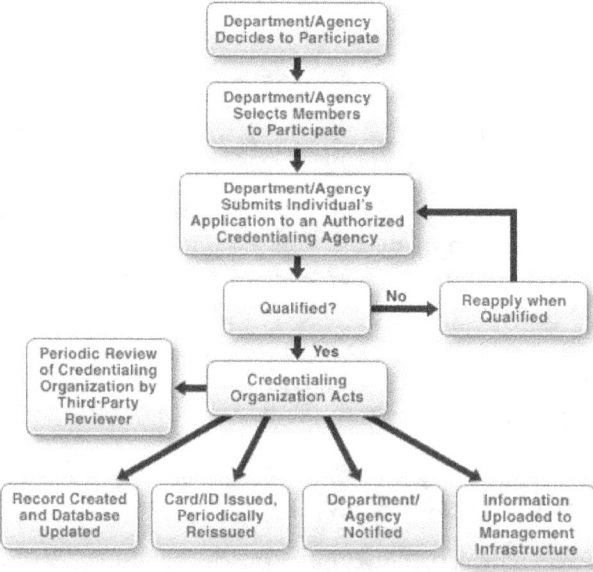

Credentialing Process

The process begins with the department/agency deciding to participate in the credentialing effort. Next the department/ agency selects members to participate in the credentialing effort.

The department/agency submits each individual's application to an authorized credentialing agency. That credentialing agency determines if the individual is qualified for the applied-for credential(s).

If the individual is found not qualified, he/she can reapply when qualified.

If the individual is found qualified, the credentialing agency acts as follows:

- Creates a record and updates the database.

- Issues a card/ID (and periodically reissues the card/ID as appropriate).

- Notifies the department/agency.

- Uploads the information to the management infrastructure.

The credentialing organization undergoes periodic review by a third-party reviewer.

Lesson 5: NIMS Command and Management

Related NIMS Document Section

This lesson summarizes the information presented in Component IV: Command and Management, including:

- Incident Command System

- Multiagency Coordination Systems

- Public Information

- Relationships Among Command and Management Elements

What Is NIMS Command and Management?

The NIMS components of Preparedness, Communications and Information Management, and Resource Management provide a framework for effective management during incident response. Next, we'll cover the fundamental elements of incident management including: Incident Command System, Multiagency Coordination Systems, and Public Information. Together, these elements comprise the NIMS Command and Management component.

The Incident Command System, or ICS, is a standardized, on-scene, all-hazard incident management concept. ICS allows its users to adopt an integrated organizational structure to match the complexities and demands of incidents.

As an incident becomes more complex, multiagency coordination becomes increasingly important. Multiagency coordination is a process that allows all levels of government and all disciplines to work together more efficiently and effectively. Multiagency coordination is accomplished through a comprehensive system of elements. These elements include facilities, equipment, personnel, procedures, and communications. Emergency Operations Centers and Multiagency Coordination Groups are just two examples of coordination elements.

The final Command and Management element is Public Information. Public Information includes processes, procedures, and organizational structures required to gather, verify, coordinate, and disseminate information – information that is essential for lifesaving response and community recovery.

NIMS is best summed up by Craig Fugate: ". . .When we fail to work as a team, we fail our citizens and what NIMS is is a system to provide a framework for all of the team to work together towards common goals."

Understanding Command and Coordination

This lesson presents information on command and coordination. Both elements are essential to ensuring a successful response. Remember that:

- **Command** is the act of directing, ordering, or controlling by virtue of explicit statutory, regulatory, or delegated authority at the field level.

- **Coordination** is the process of providing support to the command structure and may include incident prioritization, critical resource allocation, communications systems integration, and information exchange.

Command and Management Elements

Building upon all of the components covered in the previous lessons, the NIMS Command and Management component facilitates incident management. This component includes the following elements: Incident Command System, Multiagency Coordination Systems, and Public Information.

Incident Command System

The first Command and Management element is the Incident Command System (ICS).

This lesson reviews the key ICS concepts and terminology used within NIMS and is not a substitute for comprehensive ICS training. Additional information on ICS training requirements is available at the National Integration Center Web site.

What Is ICS?

ICS is a standardized, on-scene, all-hazards incident management approach that:

- Allows for the integration of facilities, equipment, personnel, procedures, and communications operating within a **common organizational structure**.

- Enables a coordinated response among various jurisdictions and functional agencies, both public and private.

- Establishes common processes for planning and managing resources.

ICS: Not Just for Large-Scale Incidents

ICS is flexible and can be used for incidents of any type, scope, and complexity.

ICS allows its users to adopt an integrated organizational structure to match the complexities and demands of single or multiple incidents.

NIMS prompts the use of ICS for every incident or scheduled event. Using ICS on all incidents helps hone and maintain skills needed for the large-scale incidents.

Management Characteristics

ICS is based on 14 proven management characteristics that contribute to the strength and efficiency of the overall system.

- Common Terminology

- Modular Organization

- Management by Objectives

- Incident Action Planning

- Manageable Span of Control

- Incident Facilities and Locations

- Comprehensive Resource Management

- Integrated Communications

- Establishment and Transfer of Command

- Chain of Command and Unity of Command

- Unified Command

- Accountability

- Dispatch/Deployment

- Information and Intelligence Management

Common Terminology

ICS establishes common terminology that allows diverse incident management and support organizations to work together across a wide variety of incident management functions and hazard scenarios. This common terminology covers the following:

- **Organizational Functions:** Major functions and functional units with incident management responsibilities are named and defined. Terminology for the organizational elements is standard and consistent.

- **Resource Descriptions:** Major resources—including personnel, facilities, and major equipment and supply items—that support incident management activities are given common names and are "typed" with respect to their capabilities, to help avoid confusion and to enhance interoperability.

- **Incident Facilities:** Common terminology is used to designate the facilities in the vicinity of the incident area that will be used during the course of the incident.

Incident response communications (during exercises and actual incidents) should feature plain language commands so they will be able to function in a multijurisdiction environment. Field manuals and training should be revised to reflect the plain language standard.

Modular Organization

The ICS organizational structure develops in a modular fashion based on the size and complexity of the incident, as well as the specifics of the hazard environment created by the incident. When needed, separate functional elements can be established, each of which may be further subdivided to enhance internal organizational management and external coordination. Responsibility for the establishment and expansion of the ICS modular organization ultimately rests with Incident Command, which bases the ICS organization on the requirements of the situation. As incident complexity increases, the organization expands from the top down as functional responsibilities are delegated. Concurrently with structural expansion, the number of management and supervisory positions expands to address the requirements of the incident adequately.

Management by Objectives

Management by objectives is communicated throughout the entire ICS organization and includes:

- Establishing overarching incident objectives.

- Developing strategies based on overarching incident objectives.

- Developing and issuing assignments, plans, procedures, and protocols.

- Establishing specific, measurable tactics or tasks for various incident management functional activities, and directing efforts to accomplish them, in support of defined strategies.

- Documenting results to measure performance and facilitate corrective actions.

Incident Action Planning

Centralized, coordinated incident action planning should guide all response activities. An Incident Action Plan (IAP) provides a concise, coherent means of capturing and communicating the overall incident priorities, objectives, and strategies in the contexts of both operational and support activities. Every incident must have an action plan. However, not all incidents require written plans. The need for written plans and attachments is based on the requirements of the incident and the decision of the Incident Commander or Unified Command. Most initial response operations are not captured with a formal IAP. However, if an incident is likely to extend beyond one operational period, become more complex, or involve multiple jurisdictions and/or agencies, preparing a written IAP will become increasingly important to maintain effective, efficient, and safe operations.

Manageable Span of Control

Span of control is key to effective and efficient incident management. Supervisors must be able to adequately supervise and control their subordinates, as well as communicate with and manage all resources under their supervision. In ICS, the span of control of any individual with incident management supervisory responsibility should range from 3 to 7 subordinates, with 5 being optimal. During a large-scale law enforcement operation, 8 to 10 subordinates may be optimal. The type of incident, nature of the task, hazards and safety factors, and distances between personnel and resources all influence span-of-control considerations.

Incident Facilities and Locations

Various types of operational support facilities are established in the vicinity of an incident, depending on its size and complexity, to accomplish a variety of purposes. The Incident Command will direct the identification and location of facilities based on the requirements of the situation. Typical designated facilities include Incident Command Posts, Bases, Camps, Staging Areas, mass casualty triage areas, point-of-distribution sites, and others as required.

Comprehensive Resource Management

Maintaining an accurate and up-to-date picture of resource utilization is a critical component of incident management and emergency response. Resources to be identified in this way include personnel, teams, equipment, supplies, and facilities available or potentially available for assignment or allocation. Resource management is described in detail in Component III.

Integrated Communications

Incident communications are facilitated through the development and use of a common communications plan and interoperable communications processes and architectures. The ICS 205 form is available to assist in developing a common communications plan. This integrated approach links the operational and support units of the various agencies involved and is necessary to maintain communications connectivity and discipline and to enable common situational awareness and interaction. Preparedness planning should address the equipment, systems, and protocols necessary to achieve integrated voice and data communications.

Establishment and Transfer of Command

The command function must be clearly established from the beginning of incident operations. The agency with primary jurisdictional authority over the incident designates the individual at the scene responsible for establishing command. When command is transferred, the process must include a briefing that captures all essential information for continuing safe and effective operations.

Chain of Command and Unity of Command

- **Chain of Command:** Chain of command refers to the orderly line of authority within the ranks of the incident management organization.

- **Unity of Command:** Unity of command means that all individuals have a designated supervisor to whom they report at the scene of the incident.

These principles clarify reporting relationships and eliminate the confusion caused by multiple, conflicting directives. Incident managers at all levels must be able to direct the actions of all personnel under their supervision.

Unified Command

In incidents involving multiple jurisdictions, a single jurisdiction with multiagency involvement, or multiple jurisdictions with multiagency involvement, Unified Command allows agencies with different legal, geographic, and functional authorities and responsibilities to work together effectively without affecting individual agency authority, responsibility, or accountability.

Accountability

Effective accountability of resources at all jurisdictional levels and within individual functional areas during incident operations is essential. Adherence to the following ICS principles and processes helps to ensure accountability:

- Resource Check-In/Check-Out Procedures

- Incident Action Planning

- Unity of Command

- Personal Responsibility

- Span of Control

- Resource Tracking

Dispatch/Deployment

Resources should respond only when requested or when dispatched by an appropriate authority through established resource management systems. Resources not requested must refrain from spontaneous deployment to avoid overburdening the recipient and compounding accountability challenges.

Information and Intelligence Management

The incident management organization must establish a process for gathering, analyzing, assessing, sharing, and managing incident-related information and intelligence.

Incident Commander

When an incident occurs within a single jurisdiction and there is no jurisdictional or functional agency overlap, a single **Incident Commander** is designated with overall incident management responsibility by the appropriate jurisdictional authority.

The designated Incident Commander develops the incident objectives that direct all subsequent incident action planning. The Incident Commander approves the Incident Action Plan and the resources to be ordered or released.

Incident Commander Responsibilities

The Incident Commander is the individual responsible for all incident activities, including the development of strategies and tactics and the ordering and the release of resources. The Incident Commander has overall authority and responsibility for conducting incident operations and is responsible for the management of all incident operations at the incident site.

The Incident Commander must:

- Have clear authority and know agency policy.

- Ensure incident safety.

- Establish the Incident Command Post.

- Set priorities, and determine incident objectives and strategies to be followed.

- Establish the Incident Command System organization needed to manage the incident.

- Approve the Incident Action Plan.

- Coordinate Command and General Staff activities.

- Approve resource requests and use of volunteers and auxiliary personnel.

- Order demobilization as needed.

- Ensure after-action reports are completed.

- Authorize information released to the media.

Unified Command

As an incident expands in complexity, **Unified Command** may be established. In a Unified Command, individuals designated by their jurisdictional or organizational authorities (or by departments within a single jurisdiction) work together to:

- Determine objectives, strategies, plans, resource allocations, and priorities.

- Execute integrated incident operations and maximize the use of assigned resources.

Advantages of Using Unified Command

In multijurisdictional or multiagency incident management, Unified Command offers the following advantages:

- A single set of objectives is developed for the entire incident.

- A collective "team" approach is used to develop strategies to achieve incident objectives.

- Information flow and coordination are improved between all jurisdictions and agencies involved in the incident.

- All agencies with responsibility for the incident have an understanding of joint priorities and restrictions.

- No agency's legal authorities are compromised or neglected.

- The combined efforts of all agencies are optimized as they perform their respective assignments under a single Incident Action Plan.

Area Command

Area Command is an organization to oversee the management of multiple incidents handled individually by separate ICS organizations.

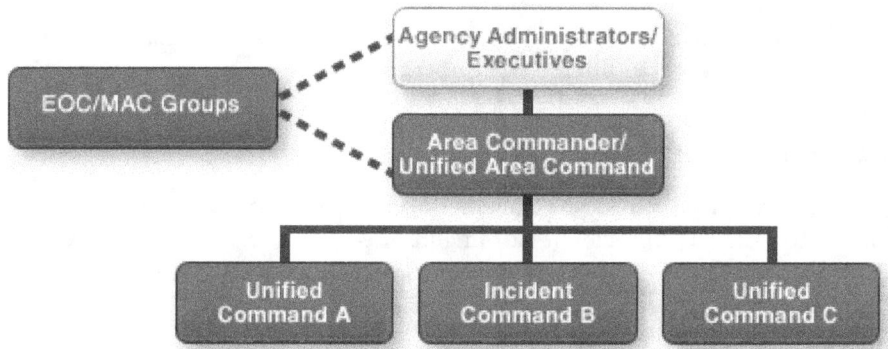

An Area Command is activated **only if necessary**, depending on the complexity of the incident and incident management span-of-control considerations.

Area Commands are particularly beneficial to incidents that are typically not site specific, are not immediately identifiable, are geographically dispersed, and evolve over longer periods of time (e.g., public health emergencies, earthquakes, tornadoes, civil disturbances). Incidents such as these, as well as acts of biological, chemical, radiological, and nuclear terrorism, require a coordinated intergovernmental, nongovernmental, and private-sector response, with large-scale coordination typically conducted at a higher jurisdictional level. Area Command is also used when a number of incidents of the same type in the same area are competing for the same resources, such as multiple hazardous material spills or fires.

For incidents under its authority, an Area Command has the following responsibilities:

- Develop broad objectives for the impacted area(s).

- Coordinate the development of individual incident objectives and strategies.

- Allocate/reallocate resources as the established priorities change.

- Ensure that incidents are properly managed.

- Ensure effective communications.

- Ensure that incident management objectives are met and do not conflict with each other or with agency policies.

- Identify critical resource needs and report them to the established EOC/MAC Groups.

- Ensure that short-term "emergency" recovery is coordinated to assist in the transition to full recovery operations.

Incident Command Post

The incident Command and Management organization is located at the Incident Command Post (ICP). Incident Command directs operations from the ICP, which is generally located at or in the immediate vicinity of the incident site. Typically, one ICP is established for each incident.

As emergency management/response personnel deploy, they must, regardless of agency affiliation, report to and check in at the designated location and receive an assignment in accordance with the established procedures.

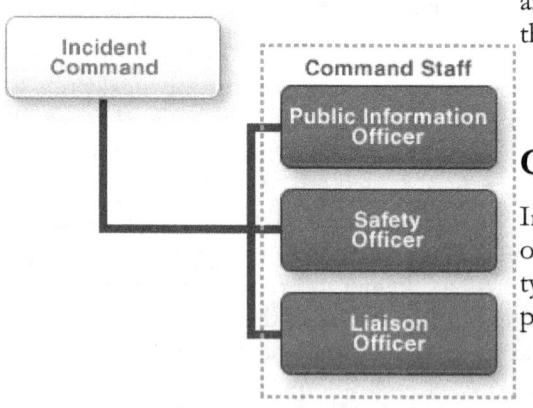

Command Staff

In an Incident Command organization, the Command Staff typically includes the following personnel:

- The **Public Information Officer** is responsible for interfacing with the public and media and/or with other agencies with incident-related information requirements.

- The **Safety Officer** monitors incident operations and advises the Incident Commander/Unified Command on all matters relating to operational safety, including the health and safety of emergency responder personnel.

- The **Liaison Officer** is the point of contact for representatives of other governmental agencies, nongovernmental organizations, and the private sector.

Additional Command Staff positions may be added depending upon incident needs and requirements.

Public Information Officer	The Public Information Officer is responsible for interfacing with the public and media and/or with other agencies with incident-related information requirements. The Public Information Officer gathers, verifies, coordinates, and disseminates accurate, accessible, and timely information on the incident's cause, size, and current situation; resources committed; and other matters of general interest for both internal and external audiences. The Public Information Officer may also perform a key public information-monitoring role. Whether the command structure is single or unified, only one Public Information Officer should be designated per incident. Assistants may be assigned from other involved agencies, departments, or organizations. The Incident Commander/Unified Command must approve the release of all incident-related information. In large-scale incidents or where multiple command posts are established, the Public Information Officer should participate in or lead the Joint Information Center in order to ensure consistency in the provision of information to the public.
Safety Officer	The Safety Officer monitors incident operations and advises the Incident Commander/Unified Command on all matters relating to operational safety, including the health and safety of emergency responder personnel. The ultimate responsibility for the safe conduct of incident management operations rests with the Incident Commander/Unified Command and supervisors at all levels of incident management. The Safety Officer is, in turn, responsible to the Incident Commander/Unified Command for the systems and procedures necessary to ensure ongoing assessment of hazardous environments, including the incident Safety Plan, coordination of multiagency safety efforts, and implementation of measures to promote emergency responder safety, as well as the general safety of

	incident operations. The Safety Officer has immediate authority to stop and/or prevent unsafe acts during incident operations. It is important to note that the agencies, organizations, or jurisdictions that contribute to joint safety management efforts do not lose their individual identities or responsibility for their own programs, policies, and personnel. Rather, each contributes to the overall effort to protect all responder personnel involved in incident operations.
Liaison Officer	The Liaison Officer is Incident Command's point of contact for representatives of other governmental agencies, nongovernmental organizations, and the private sector (with no jurisdiction or legal authority) to provide input on their agency's policies, resource availability, and other incident-related matters. Under either a single Incident Commander or a Unified Command structure, representatives from assisting or cooperating agencies and organizations coordinate through the Liaison Officer. Agency and organizational representatives assigned to an incident must have the authority to speak for their parent agencies or organizations on all matters, following appropriate consultations with their agency leadership. Assistants and personnel from other agencies or organizations (public or private) involved in incident management activities may be assigned to the Liaison Officer to facilitate coordination.
Technical Specialists	Technical specialists can be used to fill other or additional Command Staff positions required based on the nature and location(s) of the incident or specific requirements established by Incident Command. For example, a legal counsel might be assigned to the Planning Section as a technical specialist or directly to the Command Staff to advise Incident Command on legal matters, such as emergency proclamations, the legality of evacuation orders, and legal rights and restrictions pertaining to media access. Similarly, a medical advisor—an agency operational medical director or assigned physician—might be designated to provide advice and recommendations to Incident Command about medical and mental health services, mass casualty, acute care, vector control, epidemiology, or mass prophylaxis considerations, particularly in the response to a bioterrorism incident. In addition, a Special Needs Advisor

| | might be designated to provide expertise regarding communication, transportation, supervision, and essential services for diverse populations in the affected area. |

General Staff (Section Chiefs)

The General Staff includes a group of incident management personnel organized according to function and reporting to the Incident Commander. Typically, the General Staff consists of the Operations Section Chief, Planning Section Chief, Logistics Section Chief, and Finance/Administration Section Chief.

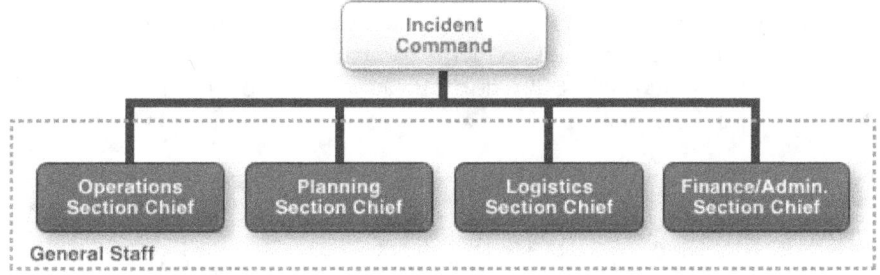

Operations Section

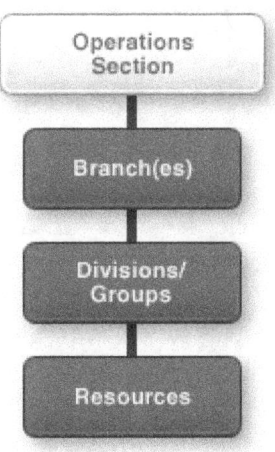

The Operations Section is responsible for all tactical activities focused on reducing the immediate hazard, saving lives and property, establishing situational control, and restoring normal operations. Lifesaving and responder safety will always be the highest priorities and the first objectives in the Incident Action Plan.

The chart on the right depicts the organizational template for an Operations Section.

Expansions of this basic structure may vary according to numerous considerations and operational factors. In some cases, a strictly functional approach may be used. In other cases, the organizational structure will be determined by geographical/jurisdictional boundaries. In still others, a mix of functional and geographical considerations may be appropriate. The ICS offers flexibility in determining the right structural approach for the specific circumstances of the incident at hand.

Operations Section Chief: The Section Chief is responsible to Incident Command for the direct management of all incident-related tactical activities.

The Operations Section Chief will establish tactics for the assigned operational period. An Operations Section Chief should be designated for each operational period, and responsibilities include direct involvement in development of the Incident Action Plan.

Branches: Branches may serve several purposes and may be functional, geographic, or both, depending on the circumstances of the incident. In general, Branches are established when the number of Divisions or Groups exceeds the recommended span of control. Branches are identified by the use of Roman numerals or by functional area.

Divisions and Groups: Divisions and/or Groups are established when the number of resources exceeds the manageable span of control of Incident Command and the Operations Section Chief. Divisions are established to divide an incident into physical or geographical areas of operation. Groups are established to divide the incident into functional areas of operation. For certain types of incidents, for example, Incident Command may assign evacuation or mass care responsibilities to a functional group in the Operations Section. Additional levels of supervision may also exist below the Division or Group level.

Resources: Resources may be organized and managed in three different ways, depending on the requirements of the incident:

- **Single Resources:** These are individual personnel, supplies, or equipment and any associated operators.

- **Task Forces:** These are any combination of resources assembled in support of a specific mission or operational need. All resource elements within a Task Force must have common communications and a designated leader.

- **Strike Teams:** These are a set number of resources of the same kind and type that have an established minimum number of personnel. All resource elements within a Strike Team must have common communications and a designated leader.

The use of Task Forces and Strike Teams is encouraged wherever possible to optimize the use of resources, reduce the span of control over a large number of single resources, and reduce the complexity of incident management coordination and communications.

Planning Section

The Planning Section collects, evaluates, and disseminates incident situation information and intelligence for the Incident Commander/Unified Command and incident management personnel. This Section then prepares status reports, displays situation information, maintains the status of resources assigned to the incident, and prepares and documents the Incident Action Plan, based on

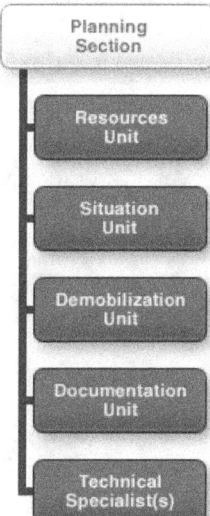

Oper ... ind guidance from the Incident Commander/Unified
Comı ...

As shown in the chart on the right, the Planning Section is comprised of four primary units, as well as a number of technical specialists to assist in evaluating the situation, developing planning options, and forecasting requirements for additional resources. These primary units that fulfill functional requirements are:

- **Resources Unit:** Responsible for recording the status of resources committed to the incident. This unit also evaluates resources committed currently to the incident, the effects additional responding resources will have on the incident, and anticipated resource needs.

- **Situation Unit:** Responsible for the collection, organization, and analysis of incident status information, and for analysis of the situation as it progresses.

- **Demobilization Unit:** Responsible for ensuring orderly, safe, and efficient demobilization of incident resources.

- **Documentation Unit:** Responsible for collecting, recording, and safeguarding all documents relevant to the incident.

- **Technical Specialist(s):** Personnel with special skills that can be used anywhere within the ICS organization.

The Planning Section is normally responsible for gathering and disseminating information and intelligence critical to the incident, unless the Incident Commander/Unified Command places this function elsewhere. The Planning Section is also responsible for assembling and documenting the Incident Action Plan.

The Incident Action Plan includes the overall incident objectives and strategies established by Incident Command. In the case of Unified Command, the Incident Action Plan must adequately address the mission and policy needs of each jurisdictional agency, as well as interaction between jurisdictions, functional agencies, and private organizations. The Incident Action Plan also addresses tactics and support activities required for one operational period, generally 12 to 24 hours.

The Incident Action Plan should incorporate changes in strategies and tactics based on lessons learned during earlier operational periods. A written Incident Action Plan is especially important when: resources from multiple agencies

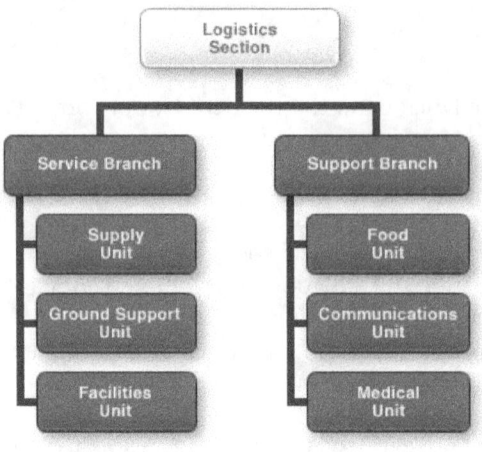

and/or jurisdictions are involved; the incident will span several operational periods; changes in shifts of personnel and/or equipment are required; or there is a need to document actions and decisions.

Logistics Section

The Logistics Section is responsible for all service support requirements needed to facilitate effective and efficient incident management, including ordering resources from off-incident locations. This Section also provides facilities, security (of the Incident Command facilities), transportation, supplies, equipment maintenance and fuel, food services, communications and information technology support, and emergency responder medical services, including inoculations, as required.

The Logistics Section is led by a Section Chief, who may also have one or more deputies. Having a deputy is encouraged when all designated units are established at an incident site. When the incident is very large or requires a number of facilities with large numbers of equipment, the Logistics Section can be divided into two Branches. This helps with span of control by providing more effective supervision and coordination among the individual units. Conversely, in smaller incidents or when fewer resources are needed, a Branch configuration may be used to combine the task assignments of individual units.

As shown in the chart on the right, the Logistics Section has six primary units that fulfill the functional requirements:

- **Supply Unit:** Orders, receives, stores, and processes all incident-related resources, personnel, and supplies.

- **Ground Support Unit:** Provides all ground transportation during an incident. In conjunction with providing transportation, the unit is also responsible for maintaining and supplying vehicles, keeping usage records, and developing incident traffic plans.

- **Facilities Unit:** Sets up, maintains, and demobilizes all facilities used in support of incident operations. The unit also provides facility maintenance and security services required to support incident operations.

- **Food Unit:** Determines food and water requirements, plans menus, orders food, provides cooking facilities, cooks, serves, maintains food service areas, and manages food security and safety concerns.

- **Communications Unit:** Major responsibilities include effective communications planning as well as acquiring, setting up, maintaining, and accounting for communications equipment.

- **Medical Unit:** Responsible for the effective and efficient provision of medical services to incident personnel.

Finance/Administration Section

A Finance/Administration Section is established when the incident management activities require on-scene or incident-specific finance and other

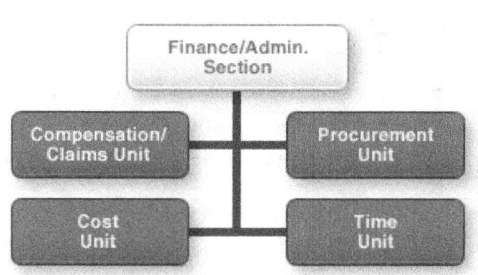

administrative support services. Some of the functions that fall within the scope of this Section are recording personnel time, maintaining vendor contracts, compensation and claims, and conducting an overall cost analysis for the incident. If a separate Finance/Administration Section is established, close coordination with the Planning Section and Logistics Section is also essential so that operational records can be reconciled with financial documents.

The Finance/Administration Section is a critical part of ICS in large, complex incidents involving significant funding originating from multiple sources. In addition to monitoring multiple sources of funds, the Section Chief must track and report to Incident Command the accrued cost as the incident progresses. This allows the Incident Commander/Unified Command to forecast the need for additional funds before operations are negatively affected.

The basic organizational structure for a Finance/Administration Section is shown in the figure on the right. Within the Finance/Administration Section, four primary units fulfill functional requirements:

- **Compensation/Claims Unit:** Responsible for financial concerns resulting from property damage, injuries, or fatalities at the incident.

- **Cost Unit:** Responsible for tracking costs, analyzing cost data, making estimates, and recommending cost-saving measures.

- **Procurement Unit:** Responsible for financial matters concerning vendor contracts.

- **Time Unit:** Responsible for recording time for incident personnel and hired equipment.

Incident Management Teams

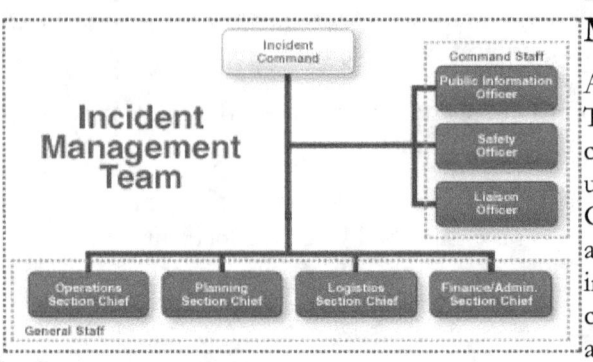

An Incident Management Team (IMT) is an incident command organization made up of the Command and General Staff members and appropriate functional units in an ICS organization and can be deployed or activated, as needed.

National, State, and some local IMTs have formal certification and qualification, notification, deployment, and operational procedures in place. In other cases, IMTs are formed at an incident or for specific events.

Multiagency Coordination Systems

The second Command and Management element is Multiagency Coordination Systems.

Multiagency coordination is a **process** that allows all levels of government and all disciplines to work together more efficiently and effectively.

The ICS 400 Advanced Incident Command System (ICS) course presents more detailed training on Multiagency Coordination Systems.

A System . . . Not a Facility

A Multiagency Coordination System is not simply a physical location or facility. Rather, a Multiagency Coordination System is a process that:

- Defines business practices, standard operating procedures, processes, and protocols by which participating agencies will coordinate their interactions.

- Provides support, coordination, and assistance with policy-level decisions to the ICS structure managing an incident.

Examples of System Elements

Multiagency coordination provides critical resource and information analysis support to the Incident Command/Unified Command. Coordination does **not** mean assuming command of the incident scene. Common coordination elements may include:

- **Dispatch Center:** A Dispatch Center coordinates the acquisition, mobilization, and movement of resources as ordered by the Incident Command/Unified Command.

- **Emergency Operations Center (EOC):** During an escalating incident, an EOC supports the on-scene response by relieving the burden of external coordination and securing additional resources. EOC core functions include coordination; communications; resource allocation and tracking; and information collection, analysis, and dissemination. EOCs may be staffed by personnel representing multiple jurisdictions and functional disciplines and a wide variety of resources.

- **Department Operations Center (DOC):** A DOC coordinates an internal agency incident management and response. A DOC is linked to and, in most cases, physically represented in the EOC by authorized agent(s) for the department or agency.

- **Multiagency Coordination (MAC) Group:** A MAC Group is comprised of administrators/executives, or their designees, who are authorized to represent or commit agency resources and funds. MAC Groups may also be known as multiagency committees or emergency management committees. A MAC Group does not have any direct incident involvement and will often be located some distance from the incident site(s) or may even function virtually. A MAC Group may require a support organization for its own logistics and documentation needs; to manage incident-related decision support information such as tracking critical resources, situation status, and intelligence or investigative information; and to provide public information to the news media and public. The number and skills of its personnel will vary by incident complexity, activity levels, needs of the MAC Group, and other factors identified through agreements or by preparedness organizations. A MAC Group may be established at any level (e.g., national, State, or local) or within any discipline (e.g., emergency management, public health, critical infrastructure, or private sector).

On-Scene and Off-Scene Multiagency Coordination

Initially the Incident Command/Unified Command and the Liaison Officer may be able to provide all needed multiagency coordination at the scene. However, as the incident grows in size and complexity, off-site support and coordination may be required.

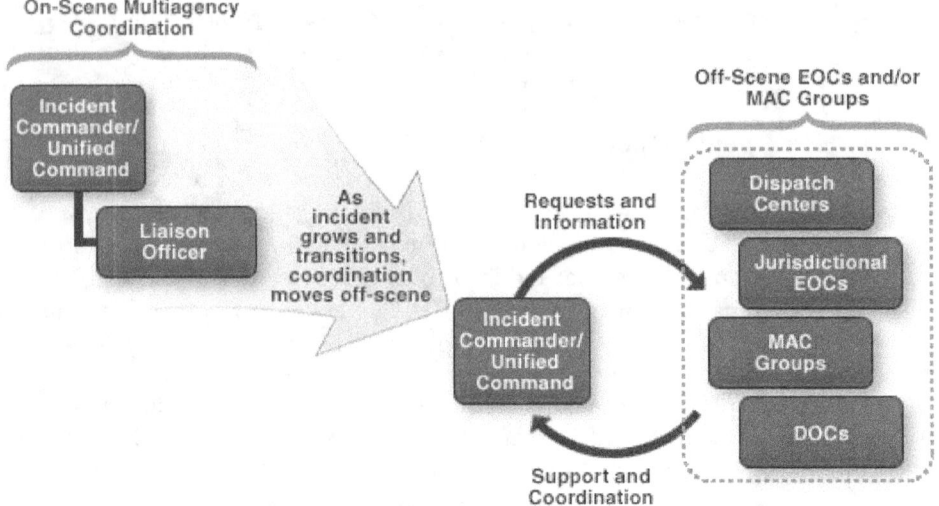

Public Information

The final Command and Management element is Public Information.

Public Information consists of the processes, procedures, and systems to communicate timely, accurate, and accessible information on the incident's cause, size, and current situation to the public, responders, and additional stakeholders (both directly affected and indirectly affected).

Public Information must be coordinated and integrated across jurisdictions, agencies, and organizations; among Federal, State, tribal, and local governments; and with nongovernmental organizations and the private sector.

Public Information

Public information, education strategies, and communications plans help ensure that numerous audiences receive timely, consistent messages about:

- Lifesaving measures.

- Evacuation routes.

- Threat and alert system notices.

- Other public safety information.

Public Information Officer

The Public Information Officer supports the incident command structure as a member of the Command Staff. Public Information Officers are able to create coordinated and consistent messages by collaborating to:

- Identify key information that needs to be communicated to the public.

- Craft messages conveying key information that are clear and easily understood by all, including those with special needs.

- Prioritize messages to ensure timely delivery of information without overwhelming the audience.

- Verify accuracy of information through appropriate channels.

- Disseminate messages using the most effective means available.

Joint Information System

The Joint Information System (JIS):

- Provides the mechanism to organize, integrate, and coordinate information to ensure timely, accurate, accessible, and consistent messaging across multiple jurisdictions and/or disciplines with nongovernmental organizations and the private sector.

- Includes the plans, protocols, procedures, and structures used to provide public information.

Federal, State, tribal, territorial, regional, or local Public Information Officers and established Joint Information Centers (JICs) are critical supporting elements of the JIS.

Joint Information Center

The Joint Information Center (JIC) is:

- A central location that facilitates operation of the Joint Information System.

- A location where personnel with public information responsibilities perform critical emergency information functions, crisis communications, and public affairs functions.

JICs may be established at various levels of government or at incident sites, or can be components of Multiagency Coordination Systems (e.g., MAC Groups or EOCs). A single JIC location is preferable, but the system is flexible and adaptable enough to accommodate virtual or multiple JIC locations, as required.

Lesson 6: Additional NIMS Elements and Resources

Related NIMS Document Section

This lesson summarizes the information presented in Component V: Ongoing Management and Maintenance, including:

- National Integration Center

- Supporting Technologies

National Integration Center

HSPD-5 required the Secretary of Homeland Security to establish a mechanism for ensuring the ongoing management and maintenance of NIMS.

The Secretary established the National Integration Center (NIC) to serve as an asset for government agencies, the private sector, and nongovernmental organizations that are implementing NIMS.

NIC Responsibilities

The NIC is responsible for the following functions:

- Administration and Compliance

- Standards and Credentialing

- Training and Exercise Support

- Publication Management

Administration and Compliance

To manage ongoing administration and implementation of NIMS, including specification of compliance measures, the NIC is responsible for working toward the following:

- Developing and maintaining a national program for NIMS education and awareness.

- Promoting compatibility between national-level standards for NIMS and those developed by other public, private, and professional groups.

- Facilitating the establishment and maintenance of a documentation and database system related to qualification, certification, and credentialing of emergency management/response personnel and organizations.

- Developing assessment criteria for the various components of NIMS, as well as compliance requirements and timelines.

Standards and Credentialing

The NIC will work with appropriate standards development organizations to ensure the adoption of common national standards and credentialing systems that are compatible and aligned with the implementation of NIMS. The standards apply to the identification, adoption, and development of common standards and credentialing programs.

Training and Exercise Support

To lead the development of training and exercises that further appropriate agencies' and organizations' knowledge, adoption, and implementation of NIMS, the NIC will coordinate with them to do the following:

- Facilitate the definition of general training requirements and the development of national-level training standards and course curricula associated with NIMS.

- Facilitate the development of national standards, guidelines, and protocols for incident management training and exercises, including consideration of existing exercise and training programs at all jurisdictional levels.

- Facilitate the development of training necessary to support the incorporation of NIMS across all jurisdictional levels.

- Establish and maintain a repository for reports and lessons learned from actual incidents, training, and exercises, as well as for best practices, model structures, and processes for NIMS-related functions.

Publication Management

Publication management for NIMS includes the development of naming and numbering conventions, the review and certification of publications, development of methods for publications control, identification of sources and suppliers for publications and related services, management of publication distribution, and assurance of product accessibility.

NIMS publication management includes the following types of products:

- Qualifications information

- Training course and exercise information

- Task books

- Incident Command System training, forms, and templates (and other necessary forms)

- Job aids and guides

- Computer programs

- Audio and video resources

- "Best practices" manuals/models/recommendations

Supporting Technologies

NIMS relies on scientifically based technical standards that support incident management. Ongoing development of science and technology supports the continual improvement and refinement of NIMS.

Strategic research and development ensures that this development takes place. To be successful, the NIC must:

- Form a long-term collaborative effort among NIMS partners to maintain an appropriate focus on science and technology solutions.

- Work in coordination with the DHS Under Secretary for Science and Technology to assess the needs of emergency management/response personnel and their affiliated organizations.

NIMS Summary

NIMS is a comprehensive nationwide framework developed through a consensus process based on incident management best practices proven by thousands of responders.

NIMS is about unifying how we respond. In time of crisis, our communities and country count on us to be able to work together as a team. We all must commit to a common way of doing business. And that way of doing business is NIMS.

www.ingramcontent.com/pod-product-compliance
Lightning Source LLC
Chambersburg PA
CBHW070837310526
45788CB00017B/1468